Functional Skills

Maths

Level 1

This incredible CGP book is ideal for Level 1 Functional Skills Maths practice.
It covers all the major test providers, including Edexcel and City & Guilds.

There are 10-Minute Tests with questions on every topic, along with tests focused
on non-calculator skills and mixed practice. We've also included easy-to-mark
answers and a progress chart to track areas where students need extra support.

CGP — still the best! ☺

Our sole aim here at CGP is to produce the highest quality books —
carefully written, immaculately presented and dangerously close to being funny.

Then we work our socks off to get them out to you
— at the cheapest possible prices.

Contents

Section Four — Mixed Practice

Published by CGP

Editors:
Michael Bushell, Tom Miles and Rosa Roberts

With thanks to Kevin Bennett and Mahjabeen Chand for the proofreading.

ISBN: 978 1 78908 484 9

Printed by Elanders Ltd, Newcastle upon Tyne.

Clipart from Corel®

Number: Test 1

There are **5 questions** in this test.
Give yourself **10 minutes** to answer them all.
You **may not** use a calculator for this test.

1   What is 76.3 × 100?

Tick (✓) your answer.

☐	0.763
☐	7.63
☐	763
✓	7630

[1]

2   What is 132 + 27 ÷ 3?

Tick (✓) your answer.

✓	53
✓	141
☐	213
☐	477

[1]

3   Use rounding to estimate the answer to 401 × 7.8.

.... 3000

[1]

4   Ranbir has been on holiday to Finland. The temperature was –5 °C
     when he left Finland and 11 °C when he arrived back in the UK.

     What was the difference in temperature between Finland and the UK?

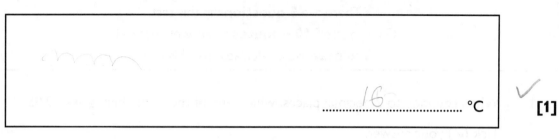

.... 16 .... °C   ✓   **[1]**

5   Rachel is buying herbs and spices from a local shop.
     She buys 37.59 g of cumin, 24.8 g of rosemary and 3.22 g of nutmeg.

     (a) What is the total weight of her herbs and spices?

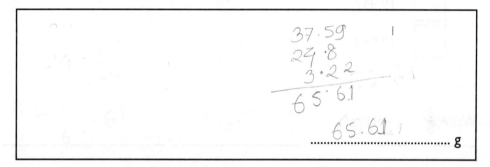

37.59
24.8
3.22
─────
65.61

.... 65.61 .... g   **[2]**

     (b) Cumin costs 4p per gram.
              How much did Rachel spend on cumin to the nearest penny?

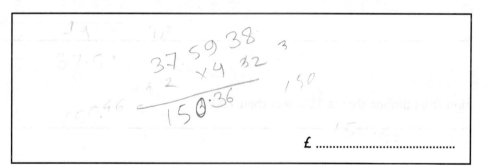

37.59 38
×4 32 3
2
─────
150.36   150

£ ....   **[2]**

END OF TEST

/ 8

Section One: Number

4

There are **5 questions** in this test.
Give yourself **10 minutes** to answer them all.
You **may** use a calculator for this test.

1 When rounded to 2 decimal places, which one of these numbers gives 19.95?

Tick (✓) your answer.

☐ 19.899

☐ 19.099

☐ 19.957

☑ 19.948 [1]

2 Write $\frac{3}{5}$ as a decimal.

0·6 [1]

3 Find the number that is 15% less than 200.

$15\% \ of \ 200 = \frac{15}{100} \times 200 = 30 \ (200-30)$

170 [1]

4   Celia needs 75% of the marks in an exam to pass. She gets $\frac{4}{5}$ of the marks.

(a) Does she pass the course? Show your working to justify your answer.

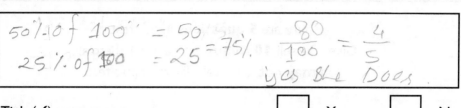

50% of 100 = 50
25% of 100 = 25 = 75% $\frac{80}{100} = \frac{4}{5}$
                         yes she Does.

Tick (✓) your answer.   ☐ Yes   ☐ No   **[1]**

(b) Her friend Hannah gets 70% of the 140 marks available in the exam.
     How many marks did she get?

70% of 140 = $\frac{70}{100} \times 140$ 98

.... marks   **[1]**

5   Elvis has two bags that each contain green and red balls.
     Each bag contains 30 balls in total.

   • In one bag, the ratio of green balls to red balls is 1 : 4.

   • In the other bag, $\frac{7}{10}$ of the balls are green.

How many out of all the balls are green?

30 ÷ 5 = 6

1 : 4   = 6 : 24
              green

$\frac{7}{10}$ of the balls are green = $\frac{21}{30}$ green

         6 + 21 = 27
.... green balls   **[3]**

END OF TEST

/ 8

Section One: Number

Number: Test 3

There are **5 questions** in this test.
Give yourself **10 minutes** to answer them all.
You **may not** use a calculator for this test.

1   Which one of the fractions below is equivalent to $\frac{2}{9}$?

Tick (✓) your answer.

☐ $\frac{9}{27}$

☐ $\frac{3}{18}$

☐ $\frac{22}{90}$

☑ $\frac{10}{45}$ **[1]**

2   What is 12.47 + 5.34 − 0.21?

12.47
5.34
17.81
.21
17.60

17·60 ....   **[1]**

3   Work out 9^2.

9 × 9 = 81

81 ....   **[1]**

4 Lucy pays £0.17 per unit of electricity.
In October, the electricity she used was worth £39.10.

How many units of electricity did she use in October?

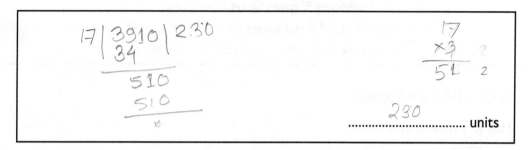

.... units **[2]**

5 A drugs company is organising a trial for a new drug.
68 people apply to take part in the trial.

(a) 50% of the people who apply are accepted. How many people is this?

68 = 50% is 34

34
.... people **[1]**

(b) Some of the people who are accepted are split between two groups in the ratio 3:2. There are 12 people in the smaller group.

How many people are in the larger group?

3 : 2
18 : 12

18
.... people **[2]**

END OF TEST

/ 8

Number: Test 4

There are **5 questions** in this test.
Give yourself **10 minutes** to answer them all.
You **may not** use a calculator for this test.

1 What is 0.45 as a fraction?

Tick (✓) your answer.

☐ $\frac{4}{10}$

☐ $\frac{7}{20}$

 $\frac{9}{20}$

☐ $\frac{7}{10}$

[1]

2 What is $6 \times (75 - 63)$?

$6 \times 12 = 72$

........................72........................ **[1]**

3 What is $\frac{1}{4}$ of 50?

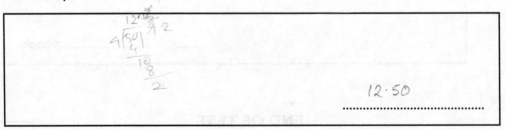

........................12·50........................ **[1]**

4   A gaming app costs £1.99 to download and offers in-app purchases
for £0.20 each. James has spent £6.59 in total on the app.

How many in-app purchases has he made?

659
199
20 | 460 | 23
     40
     60
     60

.... purchases   **[2]**

5   An artist uses the rule below to work out how much
to charge for each portrait that they paint.

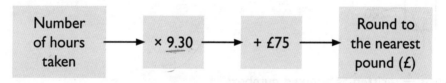

| Number of hours taken | → × 9.30 → | + £75 → | Round to the nearest pound (£) |

One portrait takes 17 hours to complete. How much will they charge for it?

17 × 93
× 717
651   2
93×
75.758·10
75
233·10

£ ....   **[3]**

END OF TEST

/ 8

Number: Test 5

There are **5 questions** in this test.
Give yourself **10 minutes** to answer them all.
You **may not** use a calculator for this test.

1 What is 80% of 300?

Tick (✓) your answer.

- [ ] 230
- [✓] 240
- [ ] 260
- [ ] 280

[1]

2 What is $\frac{7}{3}$ written as a mixed number?

Tick (✓) your answer.

- [ ] $1\frac{1}{3}$
- [✓] $2\frac{1}{3}$
- [ ] $1\frac{2}{3}$
- [ ] $2\frac{2}{3}$

[1]

3 Work out $24 - 2 \times 3$.

$2 \times 3 = 6$

$24 - 6$

18

....

[1]

4 The cost of petrol at a fuel pump is £1.28 per litre.
Marianne fills her car with the amount of petrol shown below.

How much will the petrol cost?

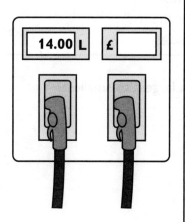

1 Liter = 1.28.
= 14 " = 1.28 × 14

128
14   3 1
———
512
28 ×
———
£7.9 2

£ .....7.9...2.......................   **[2]**

5 A charity has £58 000 to give out to deserving organisations.

£36 750 is given to a local hospital.
The remaining money is shared equally between 500 smaller organisations.

How much money does each smaller organisation receive?

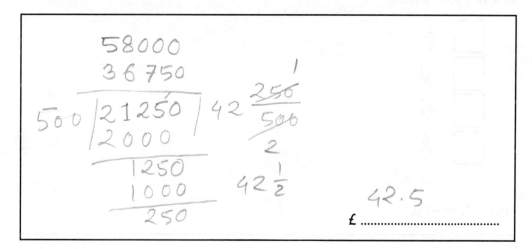

58000
36 750
——————
500 ⟌ 21250 / 42 250/500
2000
————
1250   2
1000   42 ½
————
250   42.5

£ ....   **[3]**

END OF TEST

/ 8

Number: Test 6

There are **5 questions** in this test.
Give yourself **10 minutes** to answer them all.
You **may** use a calculator for this test.

1   Which one of these sets of numbers is in order from largest to smallest?

Tick (✓) your answer.

☐	0.219	0.28	0.24	0.3
☐	0.3	0.24	0.28	0.219
✓	0.3	0.28	0.24	0.219
☐	0.219	0.24	0.28	0.3

[1]

2   Which fraction below is the biggest?

Tick (✓) your answer.

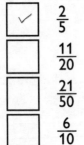

✓	$\frac{2}{5}$
☐	$\frac{11}{20}$
☐	$\frac{21}{50}$
☐	$\frac{6}{10}$

[1]

3   Find 15% of 4350.

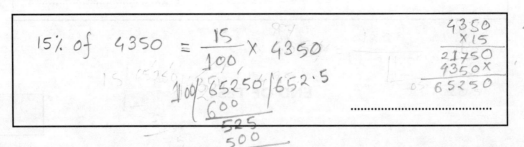

$$15\% \text{ of } 4350 = \frac{15}{100} \times 4350$$

....   **[1]**

4   Cyndi works in an office. She thinks that she spends 30 minutes in
meetings for every 3 hours that she spends working at her desk.

30

One week, she spends 3 hours and 15 minutes in meetings.

8 × 7 = 56

Using Cyndi's timings above, work out how many hours
she would have spent working at her desk.

$3 × 60 = \underline{180} + 15 = 195$ W : M

$30 \overline{)195} \, 6.5$ $\dfrac{3hr : \frac{1}{6} hr}{6 : 1}$
$\quad\,\, \underline{180}$
$\qquad 15$

$180 ÷ 30 \Big) 195$

.... hours   **[2]**

5   Usain and Kyle rented a house together for three years.
Usain paid more rent because he had a bigger room.

They always split the rent in the ratio 4 : 3.

They paid thirty thousand, two hundred and forty seven pounds
in rent over this period. How much of this did Usain pay?

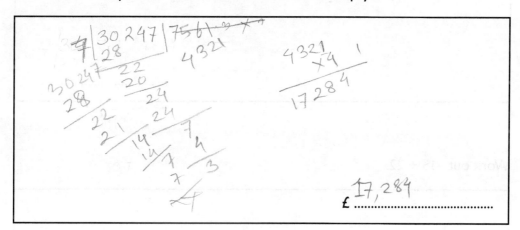

£ ....   **[3]**

END OF TEST

/ 8

Number: Test 7

There are **5 questions** in this test.
Give yourself **10 minutes** to answer them all.
You **may not** use a calculator for this test.

1   What is 34.8 ÷ 6?

Tick (✓) your answer.

$$6 \overline{)34\,8} \quad 5.8$$
$$\underline{30}$$
$$48$$
$$\underline{42}$$
$$4$$

	5.6
✓	5.7
✓	5.8
	5.9

[1]

2   What is 31.2 × 1000?

31200
....   [1]

3   Work out −15 + 22.

7
....   [1]

---

mon – Fri – 40 bu .

4 Every weekday, 40 buses go past Nino's house.
On a Sunday, only 65% of these buses operate.

How many buses go past Nino's house on a Sunday?

65 of 40 $\frac{65}{100}^{13} \times 40^{2} = 13 \times 2 = 26$

.... buses   **[2]**

26

5 Two pet rescue centres begin the year with 105 animals.
They publish these claims about how many pets they expect to rehome this year.

Centre A
We aim to have rehomed $\frac{1}{3}$ of our rescue animals.

Centre B
Our plan is for $\frac{2}{5}$ of our rescue animals to have been rehomed.

What is the difference in the number of animals that
the two centres expect to rehome this year?

Centre – A
$\frac{1}{3}$ of 105^{35} = 35   42 – 35
                                  = 7
Centre – B
$\frac{2}{5}$ of 105^{21} = 21 × 2 = 42

.... animals   **[3]**

7

END OF TEST

/ 8

Number: Test 8

There are **5 questions** in this test.
Give yourself **10 minutes** to answer them all.
You **may** use a calculator for this test.

1   What is 15^2?

Tick (✓) your answer.

☐   30

☐   125

✓   225

☐   325

[1]

2   What is $\frac{2}{7}$ of 6328?

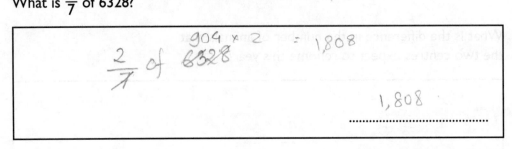

$\frac{2}{7}$ of 6328

904 × 2 = 1,808

1,808

[1]

3   What is $3\frac{1}{8}$ written as an improper fraction?

$\frac{25}{8}$

$\frac{25}{8}$

[1]

---

4   Eleanor is training for a race by running 8.5 miles each day.
She claims that she has been running more than 12 km each day.

Is she correct? Use the diagram below to check her claim. Show your working.

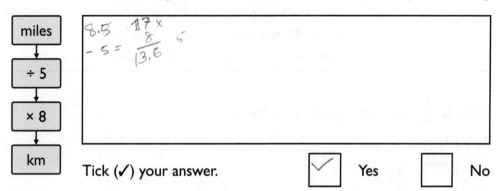

miles

÷ 5

× 8

km

8.5   17 ×
-5 =   8   5
       13.6

Tick (✓) your answer.   ✓ Yes   ☐ No   **[2]**

5   Donna buys a car on finance. The total cost of the car is split
between an initial payment and an amount to be paid over three
years in the ratio 1:5. The initial payment for the car is £1050. × 5

(a) How much does the car cost in total?

£   6300

1050
× 6
6300

6300

[2]

(b) After two years, Donna has paid off $\frac{18}{25}$ of the total cost of the car.
What is this as a percentage?   × 4

$\frac{18 \times 4}{25 \times 4}$ $\frac{72}{100}$%   3

.... %   **[1]**

END OF TEST

/ 8

Number: Test 9

There are **5 questions** in this test.
Give yourself **10 minutes** to answer them all.
You **may not** use a calculator for this test.

1 What is $\frac{8}{40}$ reduced to its simplest form?

Tick (✓) your answer.

☐ $\frac{8}{4}$

☐ $\frac{1}{8}$

☐ $\frac{1}{4}$

☑ $\frac{1}{5}$

[1]

2 What is 178 ÷ 1000?

0·178

5·6 .... [1]

3 What is 8 ÷ 0.4?

.... [1]

4 The populations of three cities are shown below.

Whenton	Whereton	Whyton
132 754	133 920	133 627

What is the total population of the two cities with the largest populations?

$$\begin{array}{r} 133920 \\ 133627 \\ \hline 267547 \end{array}$$

.... **[2]**

5 Ginny is baking mince pies.

(a) A recipe says to use 3000 g of mincemeat for 100 pies.
How many grams of mincemeat will she need for 70 pies?

100 pies for mincemeat of 3000 g

1   "   "   " $\frac{3000}{100}$ g

       "   "   " 30×700

70   ....21000.... $\frac{100}{}$ g   **[2]**

(b) Ginny has 795.817 g of mincemeat left when she has finished baking.
She splits it equally between four jars.

Estimate how much mincemeat goes into each jar.

800 g ÷ 4 jars = 200 g

.... g   **[1]**

END OF TEST

/ 8

Number: Test 10

There are **5 questions** in this test.
Give yourself **10 minutes** to answer them all.
You **may** use a calculator for this test.

1 What is 405 628 in words?

Tick (✓) your answer.

☐ four hundred and fifty thousand, six hundred and twenty-eight

✓ four hundred and five thousand, six hundred and twenty-eight

☐ forty-five thousand, six hundred and twenty-eight

☐ four million, five thousand, six hundred and twenty-eight **[1]**

2 What is the difference between the numbers labelled Y and Z on this line?

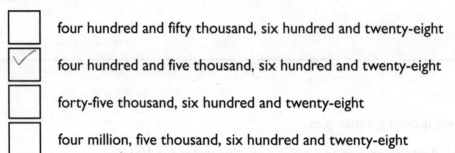

$y = -1$ $Z = 120$

$20 - -1 = 21$

.... **[1]**

3 Round 40.688888 to one decimal place.

40.7

.... **[1]**

---

4 A train company increased the price of their tickets by 5%.
A ticket from Glasgow to Lancaster before the increase cost £23.

What is the new price of the ticket?

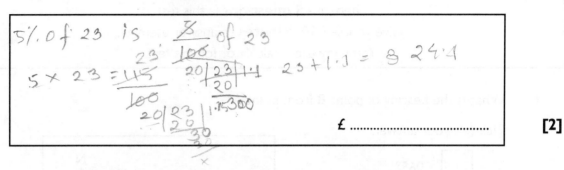

£ .... [2]

5 A business needs to move 180 000 staplers from an old building to a new one.
They use trolleys and a truck to move the staplers.

- One trolley can hold up to 600 staplers.
- The truck can hold up to 40 trolley loads.
- The journey from the old building to the new building is 12.5 miles.

The truck is currently at the old building and must return
there once all of the staplers have been moved.

How many miles will the truck have to travel to move all of the staplers?

.... miles [3]

END OF TEST

/ 8

Measures, Shape and Space: Test 1

There are **5 questions** in this test.
Give yourself **10 minutes** to answer them all.
You **may** use a calculator for this test.

1 What is the bearing of point B from point A?

Tick (✓) your answer.

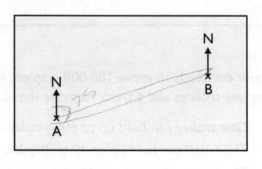

☐ 065°

☐ 285°

☐ 105°

☐ 075°   [1]

2 What shape could the plan view of a cuboid look like?

Tick (✓) your answer.

☐ circle

✓ rectangle

☐ triangle

☐ trapezium   [1]

3 Lin has a monthly income of £1389.60 and monthly living expenses of £677.80.
Last February, he spent £379.99 on a new bike.

How much income did Lin have left at the end of last February?

1389·60 138960
        −67780
         71180
       − 37999
        £ 331.81

£ ....   [1]

4 Calculate the volume of the shape below.
Give the correct units with your answer.

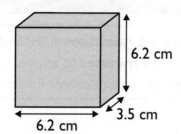

6.2 cm

3.5 cm

6.2 cm

.... **[2]**

5 Ffion and Sam are going on a hike.

(a) On their map, 5 cm represents 1 km. The path they plan to take
is 12 km long in real life. How long is this on the map?

1 km = 5 cm

12 " = 5 × 12 = 60 cm

.... cm **[1]**

(b) They plan to walk for $3\frac{1}{2}$ hours and then have a break for 35 minutes.
They will then walk for another 2 hours and 50 minutes.

If they set off at 08:30, what time would they finish their hike?

8.30   3.30 + 35 m = 4·05 ⊥ 2h 50
                              = 6.55
8·30 + 6·55 =
                              3·25

.... **[2]**

END OF TEST

/ 8

Section Two: Measures, Shape and Space

Measures, Shape and Space: Test 2

There are **5 questions** in this test.
Give yourself **10 minutes** to answer them all.
You **may not** use a calculator for this test.

1   What is 7530 g in kilograms?   7530 ÷ 1000

Tick (✓) your answer.

☐ 0.753 kg

☐ 7.53 kg

☐ 75.3 kg

☐ 753 kg   **[1]**

2   A shop offers a student discount of 10%.
How much would a student pay for a top that usually costs £38?

Tick (✓) your answer.

☐ £37.20   38 - 3·8

☐ £35.20   13.42
                                                                  9:10
✓ £34.20   4·32

☐ £28   **[1]**

3   Alun caught a train at 09:10. He got off the train at 13:42.
How long was Alun on the train for?

.... 4 ...... hours, ... 32 ...... minutes   **[1]**

Section Two: Measures, Shape and Space   © CGP — not to be photocopied

4   Jamie is thinking of a quadrilateral.

   • It has two pairs of parallel sides.
   • It has no lines of symmetry.

Draw a shape that Jamie could be thinking of on the grid below.

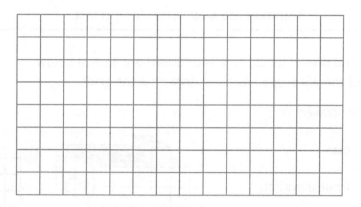

[2]

5   Nahla wants to frame a photo that has an area of 150 cm².

She uses a photo frame that has a height of 28 cm, a width of 24 cm and a wooden border of 2 cm around the perimeter.

What is the area within the wooden border that is **not** taken up by the photo?

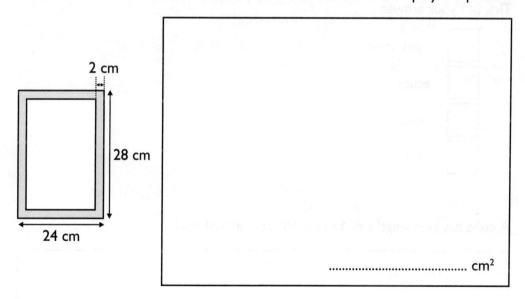

.... cm²   **[3]**

END OF TEST

/ 8

Measures, Shape and Space: Test 3

There are **5 questions** in this test.
Give yourself **10 minutes** to answer them all.
You **may** use a calculator for this test.

1   A 3D shape and one of its 2D representations are shown below.
What view of the 3D shape has been drawn on the grid?

Tick (✓) your answer.

☐ plan

☐ front elevation

☑ side elevation

Front

[1]

2   What is the name of an angle between 180° and 360°?
Tick (✓) your answer.

☐ right angle

☐ acute

☐ reflex

☐ obtuse

[1]

3   A cube has side lengths of 13 cm. What is its volume?

$13 \times 13 \times 13 =$

2,797

.... cm³   **[1]**

4   A shop has a sale of 35% off everything. Betty wants to buy
a set of plates in the sale which usually costs £28.80.

She has £17 with her. Can she afford to buy the plates? Show your working.

> 35% of 28.80 = £10.08p
> After discount = £(28.80 - 10.08)p
> = £18.72p

Tick (✓) your answer.   ☐ Yes   ☑ No   **[2]**

5   Susie is cleaning the carpet in her apartment. The carpet covers the whole
room shown below, except for the square kitchen area in the corner.

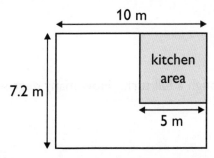

10 m

kitchen area

7.2 m

5 m

The instructions say to use 22 ml of carpet cleaner per square metre.
How many millilitres of carpet cleaner will she need?

> 7.2 × 10 = 72m − 25m = 47 m²
> 1 m² = 22 ml
> = 1,034 ml

.... ml   **[3]**

END OF TEST

/ 8

10 **Measures, Shape and Space: Test 4**

There are **5 questions** in this test.
Give yourself **10 minutes** to answer them all.
You **may** use a calculator for this test.

1   What is the name of the shape below?

Tick (✓) your answer.

☐ square

☐ trapezium

☐ rhombus

☐ parallelogram   **[1]**

2   Tommy rotates a shape through a full turn. How many degrees is this?

Tick (✓) your answer.

☐ 180°

☐ 270°

☐ 360°

☐ 720°   **[1]**

3   A map has a scale where 1 cm represents 10 m. The map shows
two buildings 7 cm apart. How far apart are the buildings in real life?

.... m   **[1]**

4 Georgia is fitting a new kitchen. The worktop has a length of 7.8 m.
The wood for the worktop costs £4.99 for every 10 cm.

How much will the wood cost for the whole of Georgia's worktop?

1 m = 1000 m = 1000 cm   1 m = 100
    7.8 m = 1000 × 7.800
          = 7800 cm

10 cm = £4.99
7800 cm = $\dfrac{4.99 \times 7800}{10}$

389.22

£ ...3,892.2... **[2]**

5 A book company wants to build a new warehouse to store its stock.
A floor plan of the warehouse is shown below.

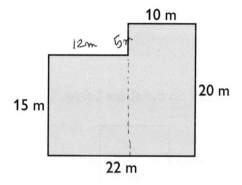

10 m

12m   5m

20 m

15 m

22 m

Work out the area of the floor.

15 m × 12 m = 180 m² }
10 m × 20 m = 200 m² } + = 380 m²

380

.... m² **[3]**

END OF TEST

/ 8

Measures, Shape and Space: Test 5

There are **5 questions** in this test.
Give yourself **10 minutes** to answer them all.
You **may** use a calculator for this test.

1   What type of triangle has three angles of the same size?

Tick (✓) your answer.

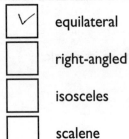

equilateral

right-angled

isosceles

scalene

[1]

2   Draw all the lines of symmetry on the shape below.

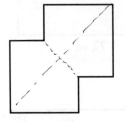

[1]

3   Katie takes out a loan of £1650 to pay for a college course.
The interest rate is 5% per year.

How much interest will she owe on this loan
after the first year if no repayments are made?

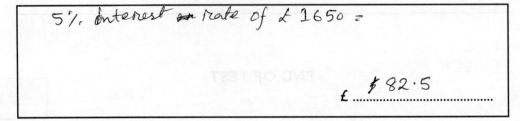

5% interest ~~at~~ rate of £1650 =

£ ...... ~~£~~ 82·5 ......   **[1]**

---

4   Sunil is selling his handmade wooden boxes at a market.
He sells each box for £19. The materials needed to make one box cost £6.24.

He has five boxes to sell. If he sells all of them, how much profit will he make?

> materials cost per box = £6.24
>
>   5   u
> He sells each box =£19 × 5 =£95   =£6.24 × 5 = 31.42
>   =£95
> (31.2 − 95) = 63.6
>
>   £ ...63.8...   **[2]**

5   Jodie fills her watering can from a cuboid water butt, shown below.
The water butt has a depth of 2 m, a width of 0.6 m and a length of 0.5 m.

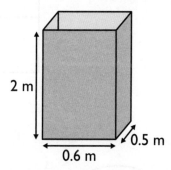

2 m

0.5 m

0.6 m

The capacity of Jodie's watering can is 5000 cm³.

If the watering butt starts off full, how many times could
Jodie fill her watering can before the butt is empty?

> 2 m × 0.6 m × 0.5 m = 0.6 m³
>
> 200 cm × 60 m × 5 m = 600,000 cm³
>
> (600,000 ÷ 5000) cm³ = 120 cm³
>
>   120
>   .... times   **[3]**

END OF TEST

/ 8

Measures, Shape and Space: Test 6

There are **5 questions** in this test.
Give yourself **10 minutes** to answer them all.
You **may** use a calculator for this test.

1 Look at the net below. What 3D shape does it belong to?

Tick (✓) your answer.

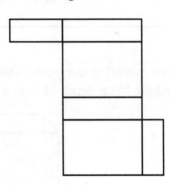

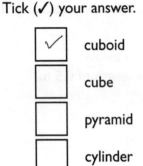

✓	cuboid
	cube
	pyramid
	cylinder

[1]

2 Draw a radius on the circle below.

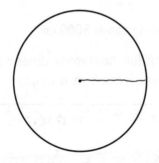

[1]

3 Jordie has a £20 gift card. She uses the card to buy 6 truffles which cost £1.82 each. What will the balance be on the card after she has bought the truffles?

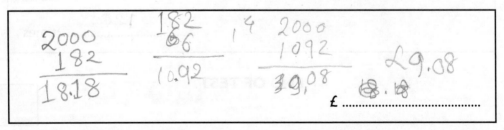

2000
182
18.18

182
×6
10.92

14
2000
1092
£9.08

2000
1092
19.08

£9.08

8.8

£ ....

[1]

---

4 Val is making greetings cards. She wants to put a ribbon around the edge of the front of every card. Each card has a length of 11.2 cm and a width of 8.7 cm.

How many centimetres of ribbon will she need for 10 cards?

.... cm **[2]**

5 The map below shows the positions of Carkmel and Moresham and the routes between them by train and by boat.

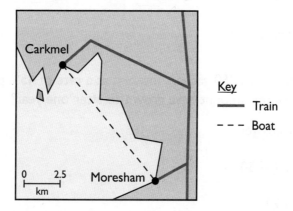

How many kilometres further is the journey by train than by boat?

.... km **[3]**

END OF TEST

/ 8

Section Two: Measures, Shape and Space

Measures, Shape and Space: Test 7

There are **5 questions** in this test.
Give yourself **10 minutes** to answer them all.
You **may** use a calculator for this test.

1   How many lines of symmetry does an isosceles triangle have?

Tick (✓) your answer.

☐ none

☐ one

☐ two

☐ three   **[1]**

2   An initial amount of £250 gathers interest at a rate of 45% per year.
What value will the amount have grown to after one year?

Tick (✓) your answer.

☐ £137.50

☐ £295

☐ £335

✓ £362.50   **[1]**

3   Work out the missing length in the shape below.

3 m

?

2 m

7 m

1.5 m

5 m

5.5 .... m   **[1]**

Section Two: Measures, Shape and Space   © CGP — not to be photocopied

4 Below is a cube with a side length of 2 cm.
Draw a net of this cube on the grid below.

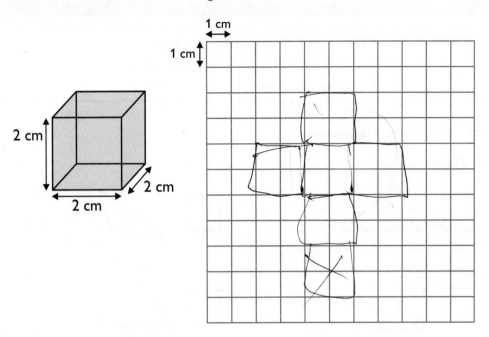

[3]

5 Yemi is doing some laundry. He uses 0.04 litres of detergent per load.
The amount of laundry detergent he has left is shown in the container below.

He has eight loads of laundry left to do. Does he have
enough detergent to do them all? Show your working.

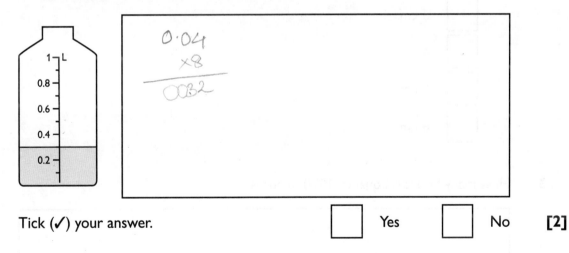

Tick (✓) your answer.   ☐ Yes   ☐ No   **[2]**

END OF TEST

/ 8

Measures, Shape and Space: Test 8

There are **5 questions** in this test.
Give yourself **10 minutes** to answer them all.
You **may** use a calculator for this test.

1   Which triangle below is scalene?

Tick (✓) your answer.

[1]

2   A cuboid has a height of 12 cm, a depth of 7 cm and a volume of 336 cm³.
What is the width of the cuboid?

Tick (✓) your answer.

- 3 cm
- 4 cm
- 5 cm
- 6 cm

[1]

3   How many hours is equal to 1920 minutes?

32 .... hours   [1]

---

4   Work out the perimeter of the shape below.

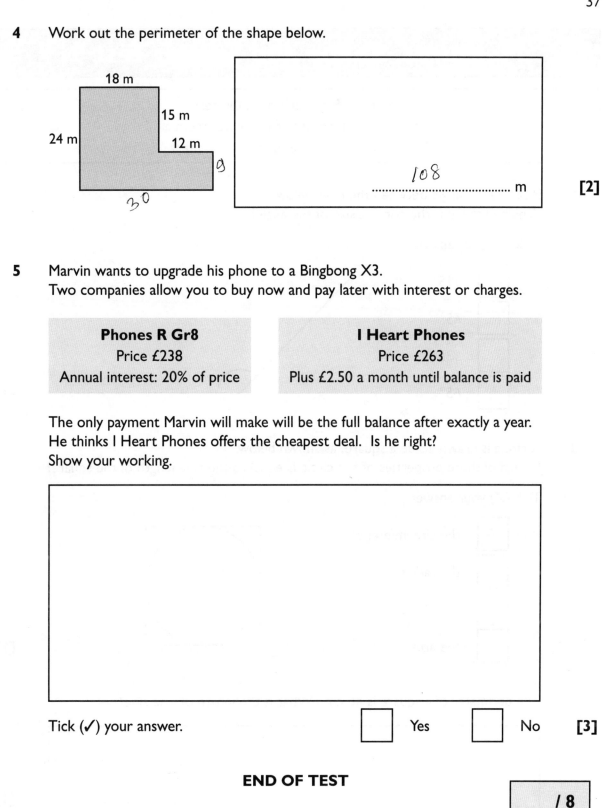

18 m

15 m

24 m

12 m

9

30

.... m   **[2]**

108

5   Marvin wants to upgrade his phone to a Bingbong X3.
Two companies allow you to buy now and pay later with interest or charges.

Phones R Gr8
Price £238
Annual interest: 20% of price

I Heart Phones
Price £263
Plus £2.50 a month until balance is paid

The only payment Marvin will make will be the full balance after exactly a year.
He thinks I Heart Phones offers the cheapest deal. Is he right?
Show your working.

Tick (✓) your answer.   ☐ Yes   ☐ No   **[3]**

END OF TEST

/ 8

Measures, Shape and Space: Test 9

There are **5 questions** in this test.
Give yourself **10 minutes** to answer them all.
You **may not** use a calculator for this test.

1   Measure the angle between the lines below.
Which of these is the correct size of the angle?

Tick (✓) your answer.

115°

45°

135°

65°   [1]

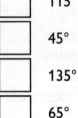

2   A circle is drawn inside a square, as shown below.
Which of these properties of the circle is equal to the side length of the square?

Tick (✓) your answer.

the circumference

the radius

the diameter

the area   [1]

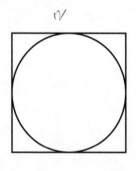

3   The perimeter of a square is 48 cm. What is the length of one side?

.... cm   [1]

---

4   Marina has £435 in her bank account. She goes to a department store and
      buys a shirt, headphones and a coffee machine, as shown on the receipt below.

      After buying these items, she has £239.25 left in her account.
      How much did the coffee machine cost?

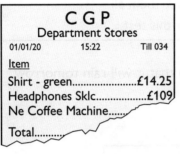

CGP
Department Stores
01/01/20   15:22   Till 034
Item
Shirt - green........................£14.25
Headphones Sklc.................£109
Ne Coffee Machine.......
Total..........

£ .... [2]

5   Roger is taking a train from York to London to watch a theatre show.

      The show starts at 7.30 pm. He wants to arrive at the theatre at least
      half an hour before the show starts. It will take him 45 minutes to travel
      from the train station in London to the theatre.

7.00
6.15

      The train timetable is shown below.
      What time does the latest train that he could catch depart York?

Timetable					
Edinburgh	1245	1300		1400	
Newcastle	1406	1451	1458	1532	1559
York	1515	1550	1600	1637	1701
Grantham	1609	1637	1704	1735	1800
London	1705	1742	1803	1831	1855

.... [3]

END OF TEST

/ 8

Handling Data: Test 1

There are **4 questions** in this test.
Give yourself **10 minutes** to answer them all.
You **may not** use a calculator for this test.

1 The arrow on the scale below shows the likelihood that it will rain tomorrow.
How would you describe this likelihood?

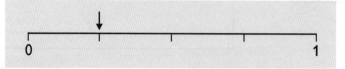

Tick (✓) your answer.

☐	Certain	☑	Unlikely
☐	Likely	☐	Impossible

[1]

2 The table below shows the number of tickets sold on
a Saturday at five different cinema chains across the UK.

Cinema chain	Number of tickets sold
Watch	168 190
Projector	239 477
Anyperson	182 056
Screenhome	161 503
Movieglobe	196 842

Work out the range of the numbers of tickets sold.

$$\begin{array}{r} 239477 \\ 161503 \\ \hline 77974 \end{array}$$

.... tickets **[2]**

3   A group of car owners were asked their age. Some of the results are in this table.

Age	0-20	21-40	41-60
Frequency	12	17	28

(a) Complete the chart below to illustrate this information.

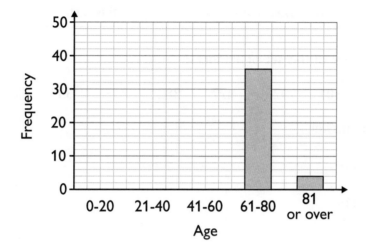

[2]

(b) How many of the people asked were at least 41 years old?

.... people   [1]

4   A lucky dip is made up of 20 prizes. Five of the prizes are boxes
of chocolates, nine are tubes of sweets and the rest are notebooks.

What is the probability of **not** picking a notebook from the lucky dip?

[2]

END OF TEST

/ 8

Handling Data: Test 2

There are **3 questions** in this test.
Give yourself **10 minutes** to answer them all.
You **may not** use a calculator for this test.

1   What is the range of this set of numbers?

| 301 | 423 | 477 | 382 | 415 | 339 |

Tick (✓) your answer.

☐   138

☐   114

☑   176

☐   166

$$\begin{array}{r} 477 \\ 301 \\ \hline 176 \end{array}$$

[1]

2   Sally counts up how many nails, screws and bolts she has in her toolbox. She puts the results in a tally chart.

(a)   Fill in the missing boxes in the tally chart on the right.

	Tally	Frequency
Nails	ⵑⵑⵑ IIII	9
Screws	ⵑⵑⵑ I	6
Bolts	IIII	
	Total:	19

[1]

(b)   She puts all the nails, screws and bolts in a tub, then carefully picks one out at random. What is the probability that she picks out a screw?

$$\frac{6}{19}$$

[1]

Section Three: Handling Data

3 A shopkeeper records the number of people who visited their shop each morning from Monday to Friday. The results are shown in the pie chart.

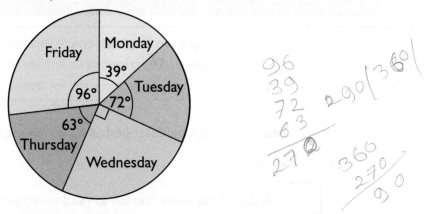

(handwritten working:)
96
39
72
63
290 ⟌ 360

270

360
270
90

(a) What percentage of the total number of visitors came on Wednesday?

(handwritten: 25)

.... %   **[1]**

(b) There were a total of 120 visitors recorded.
How many degrees represents one visitor on the pie chart?

(handwritten:) 120 ⟌ 360 ⟌ 3°

.... °   **[2]**

(c) How many **more** people visited the shop on Tuesday than on Monday?

(handwritten:) 72 − 39 = 33

....   **[2]**

END OF TEST

/ 8

Section Three: Handling Data

Handling Data: Test 3

There are **3 questions** in this test.
Give yourself **10 minutes** to answer them all.
You **may** use a calculator for this test.

1   Which of these events can be described as having an even chance of happening?

Tick (✓) your answer.

☐   A day of the week chosen at random ends in the letter 't'.

☐   A card picked at random from a set labelled 1 to 20 is odd.

☐   The next car to drive past your home is red.

☐   February will be followed by March next year.   **[1]**

2   Jan is the manager of a warehouse. She needs a table in which she can record
the number of pallets dispatched that day. She wants to group the data based
on the number of boxes on each pallet. A maximum of 20 boxes fit on a pallet.

Design a sensible table in the space below that Jan could use.
There should be at least three groups.

[2]

3   A driving school has four instructors — Pam, Ian, Jeff and Sue.
They each work out the mean number of tests their students
have taken and draw it on this bar chart.

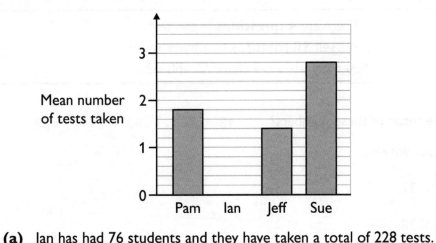

(a)   Ian has had 76 students and they have taken a total of 228 tests.
Draw the bar for Ian.   **[2]**

(b)   Jeff claims that the bar chart shows that each of his students has taken half
as many tests as Sue's students. Is Jeff correct? Explain your answer.

....

....

....   **[1]**

(c)   Pam's students have taken a total of 396 tests.
How many students has she taught?

.... students   **[2]**

END OF TEST

/ 8

Handling Data: Test 4

There are **4 questions** in this test.
Give yourself **10 minutes** to answer them all.
You **may** use a calculator for this test.

1   What is the mean of these numbers?   18   66   53   14   39

Tick (✓) your answer.

☐	37
✓	38
☐	39
☐	40

[1]

2   A sensor records the temperature at a weather station every 3 hours.
The results are shown in the table below.

Draw a line graph on the grid to represent this data.

Time	Temp. (°C)
00:00	6
03:00	5
06:00	9
09:00	13
12:00	19
15:00	22
18:00	15
21:00	10

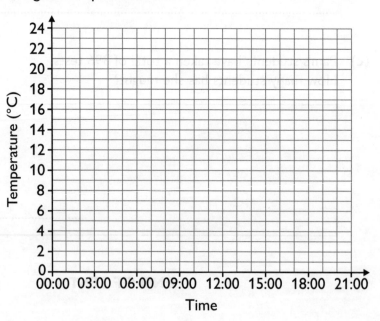

[2]

---

3   Four people were asked to rate the look, taste and texture of a new ready meal.
    They rated each quality out of 5. The table below shows their results.

Name	Look (/5)	Taste (/5)	Texture (/5)	Total (/15)
Anette	4.0	2.0	2.5	8.5
Bernard	3.1	3.7	3.4	10.2
Cate	4.0	3.9	3.5	11.4
Daveed	2.3	4.1	3.8	10.2

(a) How many more points did Bernard give for texture than Anette?

.... points   **[1]**

(b) Who had the biggest range within their own scores? Show your working.

....   **[2]**

4   Iasha rolls a 5 on a standard six-sided dice. She then rolls a second six-sided dice.
    What is the probability that the results sum to 10 or more?

$\frac{2}{6}$ [~~5~~] / [10]   **[2]**

END OF TEST

/ 8

Handling Data: Test 5

There are **3 questions** in this test.
Give yourself **10 minutes** to answer them all.
You **may** use a calculator for this test.

1   At a factory, 8 out of every 20 products made are shipped overseas.
     Which of these arrows shows the probability that a randomly selected
     product will be shipped overseas?

     Tick (✓) your answer.

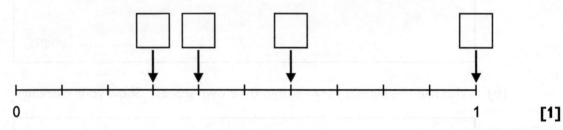

     0   1   **[1]**

2   A railway engineer is producing a report on the causes of
     train cancellations in her region last year.

Cause	Engineering works	Trespass	Signal failure	Weather
Number of cancellations	130	30	220	340

     Use the data in the table to complete this pie chart for her report.

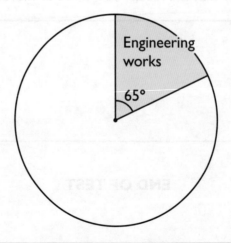

     [3]

Section Three: Handling Data

3 Sasha is researching the cost of a smoothie and a burger at a music festival in the years 2015, 2017 and 2019 compared to 2013. He produces this bar chart.

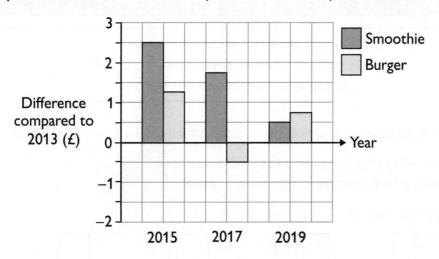

(a) How much more did a smoothie cost in 2015 than in 2013?

£ .... **[1]**

(b) What was the average difference compared to 2013 of the cost of a burger for the three years shown?

£ .... **[2]**

(c) Sasha claims that a burger was more expensive than a smoothie in 2019. Can you tell from the chart whether he is correct? Explain your answer.

....

.... **[1]**

END OF TEST

/ 8

Handling Data: Test 6

There are **4 questions** in this test.
Give yourself **10 minutes** to answer them all.
You **may** use a calculator for this test.

1 There are 10 bunches of flowers on a stall. Wendy picks a bunch at random.

The probability that she picks a bunch of roses is $\frac{1}{5}$.
How many of the bunches of flowers are roses?

Tick (✓) your answer.

☐ 1 ☐ 2 ☐ 3 ☐ 4 **[1]**

2 The number of hours Lyla spent at the gym over 3 years is shown on this graph.

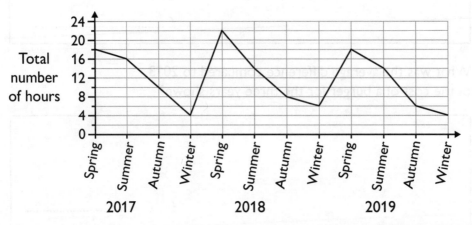

(a) What is the range of hours spent at the gym over the three year period?

.... hours **[1]**

(b) During which season does Lyla spend the most time at the gym?

.... **[1]**

3   The prices of 21 items in a supermarket are written below.
Use this data to complete the table on the right.

£7.90	£10.82	£9.22
£9.35	£10.19	£9.17
£11.00	£9.60	£8.71
£9.46	£9.90	£10.05
£8.56	£11.88	£8.59
£10.63	£8.99	£9.34
£9.72	£9.83	£8.19

Price	Frequency
£8.50 or less	
£8.51-£9.50	
£9.51-£10.50	
£10.51 or over	

[2]

4   Ken is comparing two airlines which fly to Madrid.
This table shows how many minutes of delays
Airline A has suffered in the past 5 years.

Year	Minutes delayed
2014	8220
2015	8457
2016	7950
2017	8065

(a)   Airline B had a mean of 8148 minutes of delays
over the same period. Which airline had the fewer
minutes of delays on average? Show your working.

Airline .... [2]

(b)   The range of the data for Airline B was 362 minutes during this period.
If the lowest value was 7879 minutes, what was the greatest value?

.... minutes   [1]

END OF TEST

/ 8

Mixed Practice: Test 1

There are **5 questions** in this test.
Give yourself **10 minutes** to answer them all.
You **may not** use a calculator for this test.

1   What is 1000 + 100 450 ÷ 1000?

Tick (✓) your answer.

☐	101.45
☐	110.45
☐	1010.45
☐	1100.45

[1]

2   Three people rate a new phone using proportions, as shown in the table below.

Which of these lists puts the people in order, starting with
the person who gave the highest rating?

Tick (✓) your answer.

☐	Ada, Ben, Cal
☐	Ben, Ada, Cal
☐	Ben, Cal, Ada
☐	Cal, Ada, Ben

Name	Rating
Ada	75%
Ben	$\frac{4}{5}$
Cal	0.6

[1]

3   A can of cola has a capacity of 330 ml. How much cola would fill $\frac{2}{3}$ of the can?

.... ml   **[1]**

Section Four: Mixed Practice   © CGP — not to be photocopied

4   What is the range of these numbers?

| 82.95 | 81.45 | 36.36 | 35.88 | 36.49 | 82.62 |

.... **[2]**

5   Siobhan has a rectangular yard. The perimeter of the yard is 16.8 m.

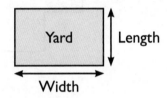

Length

Width

The yard's length and width are in the ratio 3:4.
What is the width of the yard?

.... m **[3]**

END OF TEST

/ 8

Mixed Practice: Test 2

There are **5 questions** in this test.
Give yourself **10 minutes** to answer them all.
You **may** use a calculator for this test.

1   How many lines of symmetry does the shape below have?

Tick (✓) your answer.

☐   0

☐   2

☐   4

☐   8   **[1]**

2   A boat worth £206 250 was sold at auction with a 5% discount.
What was the selling price of the boat in words?

Tick (✓) your answer.

☐   One hundred and ninety-five thousand,
    ninety-three hundred and thirty pounds and fifty pence

☐   One hundred and ninety-five thousand,
    nine hundred and thirty-seven pounds and five pence

☐   One hundred and ninety-five thousand,
    nine hundred and thirty-seven pounds and fifty pence   **[1]**

3   What fraction is equivalent to $2\frac{3}{4}$?

$$\frac{\boxed{}}{4}$$ **[1]**

4   A pet passport costs £63.98. To keep the passport, the pet must also have an injection before the passport is received and then again after three years. Each injection costs £32.60.

What is the total cost of the passport and injections over a period of 11 years?

£ ....   **[2]**

5   This table shows the number of houses listed at an estate agents over a year.

Winter			Spring		
Dec	Jan	Feb	Mar	Apr	May
8	7	9	13	17	21
Summer			Autumn		
Jun	Jul	Aug	Sep	Oct	Nov
26	24	16	14	14	11

Complete the bar chart below to show the **mean** number of houses listed in a month for each of the four seasons.

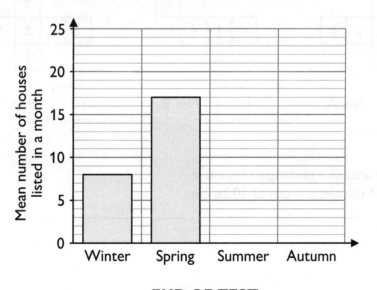

[3]

END OF TEST

/ 8

Mixed Practice: Test 3

There are **5 questions** in this test.
Give yourself **10 minutes** to answer them all.
You **may not** use a calculator for this test.

1   What fraction of this circle has been shaded?
     Use a protractor.

     Tick (✓) your answer.

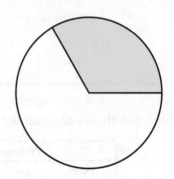

     ☐ $\frac{100}{180}$ ☐ $\frac{120}{180}$

     ☐ $\frac{60}{360}$ ☐ $\frac{120}{360}$ **[1]**

2   A dice is numbered so that opposite sides add up to seven.
     Which one of these diagrams could be the net of this dice?

     Tick (✓) your answer.

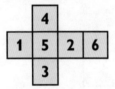   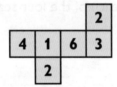   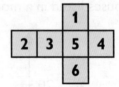

     ☐ Net A   ☐ Net B   ☐ Net C   **[1]**

3   Estimate the area of a rectangle that has a
     height of 3.94 cm and a width of 10.26 cm.

     .... cm²   **[1]**

4 A flower bed contains red roses and white roses in the ratio 2:3.
Put a cross (✗) on the scale below to show the probability
that a rose chosen at random is coloured red.

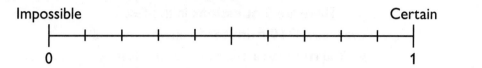

Impossible   Certain

0   1   **[1]**

5 Kelly sells bags of bird seed at a park. This box shows the
weights (in grams) of the 20 bags she sold this morning.

543	572	526	512	542	561	588	568	590	534
504	562	576	530	545	571	573	525	569	573

(a) Use this data to complete the table below.

Weight (g)	Tally	Frequency
500-524	||	2
525-549	ⅢⅡ ||	7
550-574		
574-600		
Total		20

[2]

(b) Yesterday, a bag was sold that weighed 10% more than the heaviest bag
that was sold this morning. What was the weight of yesterday's bag?

.... g   **[2]**

END OF TEST

/ 8

Mixed Practice: Test 4

There are **5 questions** in this test.
Give yourself **10 minutes** to answer them all.
You **may** use a calculator for this test.

1 What is the difference between –995 and 1000?

Tick (✓) your answer.

☐ –5

☐ 5

☐ 1995

☐ –1995 **[1]**

2 Calculate the area of the shape below.

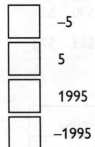

3 m

2 m

6 m

4 m

.... m² **[1]**

3 1 cm on a map represents 100 cm in real life.
A trail on the map is 12 cm long. How many **metres** is the trail in real life?

.... m **[1]**

4   Kamil borrows £1299 to refit his bathroom. He must pay back the money, plus 15% interest. How much money will he owe in total?

£ ....   **[2]**

5   A cereal company asked 60 customers whether or not they agree with the following statement:

"Yoghurt is better than milk on cereal."

The table on the right shows the results.

Response	People
Agree	25
Disagree	65
Undecided	30
Total	120

Complete the pie chart below to display this data.
You should label each sector with its angle and the response that it represents.

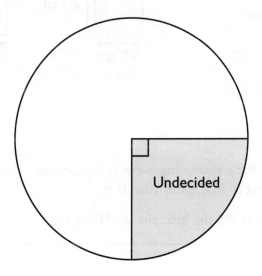

[3]

END OF TEST

/ 8

Mixed Practice: Test 5

There are **5 questions** in this test.
Give yourself **10 minutes** to answer them all.
You **may** use a calculator for this test.

1 Which of these values are smaller than $\frac{2}{5}$?

Tick (✓) **all** of your answers.

☐ $\frac{1}{2}$ ☐ $\frac{3}{10}$ ☐ $\frac{3}{4}$ ☐ $\frac{2}{8}$ **[1]**

2 The front and side elevations of a cuboid are shown below.
What is the volume of the cuboid?

Tick (✓) all of your answers.

☐ 7.935 cm³

☐ 8.97 cm³

☐ 13.754 cm³

☐ 20.631 cm³

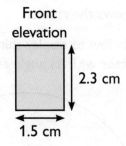

Front
elevation

2.3 cm

1.5 cm

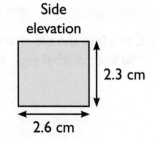

Side
elevation

2.3 cm

2.6 cm

[1]

3 The qualifying time for a swimming event is 14 minutes.
Sadia sets a time that is **30% faster** than this.

By how many minutes did she beat the qualifying time?

.... minutes **[1]**

4   A doctor's clinic opens at 9 am and closes at 1 pm.
The doctor's time is split evenly into 20 minute appointments.

Rhys books one of these appointments with the doctor at random.
What is the probability that his appointment is at 9 am?

[2]

5   A bulb lights up an area of a theatre stage.
The width of the lit area can be found using this formula:

Width of lit area in metres = 100 ÷ (height of bulb in metres + 64)

What is the width of the lit area when the bulb is 8.96 m above the stage?
Give your answer as a decimal, correct to two decimal places.

.... m   [3]

END OF TEST

/ 8

Answers

...ber

...3

1

2 $132 + 2$,
$= 132 + 9 = $ **141** *(1 mark)*
You have to do the division before the addition.

3 $401 × 7.8$
$≈ 400 × 8 = $ **3200** *(1 mark)*

4 $11 - -5 = 11 + 5$
$= $ **16 °C** *(1 mark)*

5 a)
```
   3 7 . 5 9
   2 4 . 8 0
 +   3 . 2 2
 ───────────
   6 5 . 6 1 g
   1 1 1
```
(1 mark for a correct method, 1 mark for the correct answer)

b)
```
     3 7 . 5 9
 ×           4
 ─────────────
 1 5₃0₂. 3₃6
```
150.36p = **£1.50** (to the nearest penny)
(1 mark for the correct multiplication, 1 mark for the correct answer)

Test 2 — pages 4 and 5

1 **19.948** *(1 mark)*

2 $\frac{3}{5} = \frac{6}{10} = $ **0.6** *(1 mark)*

3 15% of 200 $= \frac{15}{100} × 200 = 30$
So 15% less than 200
$= 200 - 30 = $ **170** *(1 mark)*

---

4 a) $\frac{4}{5} = 0.8 = 80\%$
Yes — she does pass. *(1 mark)*

b) 70% of 140
$= \frac{70}{100} × 140 = 98$
So she gets **98 marks**.
(1 mark)

5 In the first bag, there are
$1 + 4 = 5$ parts,
so $30 ÷ 5 = 6$ green balls.
In the second bag, there are
$\frac{7}{10} × 30 = 21$ green balls.
So across both bags,
$6 + 21 = $ **27** of the balls are green.
✱ **(1 mark for finding the number of green balls in the first bag, 1 mark for finding the number of green balls in the second bag, 1 mark for the correct answer)**

Test 3 — pages 6 and 7

1 $\frac{10}{45}$ *(1 mark)*

2
```
   1 2 . 4 7
 +   5 . 3 4
 ───────────
   1 7 . 8 1
         1
 -   0 . 2 1
 ───────────
   1 7 . 6 0
```
(1 mark)

3 $9^2 = 9 × 9 = $ **81** *(1 mark)*

4 $39.10 ÷ 0.17 = 3910 ÷ 17$:
```
       0 2 3 0
 17 | 3 9⁵1 0
```
So she used **230 units** of electricity.
(1 mark for a correct method, 1 mark for the correct answer)

---

5 a) 50% of 68
$= 68 ÷ 2$
$= $ **34 people** *(1 mark)*

b) $12 ÷ 2 = 6$ people
$3 × 6 = $ **18 people**
(1 mark for working with the proportions, 1 mark for the correct answer)

Test 4 — pages 8 and 9

1 $0.45 = \frac{45}{100} = \frac{9}{20}$ *(1 mark)*

2 $6 × (75 - 63)$
$= 6 × 12 = $ **72** *(1 mark)*

3 $\frac{1}{4}$ of $50 = 50 ÷ 4$
$= $ **12.5** *(1 mark)*
$12\frac{1}{2}$ would also be right.

4 Firstly, work out how much he spent on in-app purchases by subtracting the download cost:
```
   ⁵ ¹⁵
   6̶ . 5̶ 9
 - 1 . 9 9
 ─────────
   4 . 6 0
```
Then divide to find the number of purchases:
```
       0 2 3
 20 | 4 6⁶0
```
So he made **23** in-app purchases.
(1 mark for finding £4.60, 1 mark for the correct answer)

34

5

$$
\begin{array}{r}
1\ 7 \\
\times\ \ 9\ 3 \\
\hline
5\,_2 1 \\
+\ 1\ 5\,_6 3\ 0 \\
\hline
1\ 5\ 8\ 1
\end{array}
$$

So 17 × £9.30 = £158.10

$$
\begin{array}{r}
1\ 5\ 8\ .\ 1\ 0 \\
+\ \ \ 7\ 5\ .\ 0\ 0 \\
\hline
2\ 3\ 3\ .\ 1\ 0 \\
{}_1\ \ {}_1
\end{array}
$$

So £158.10 + £75
= £233.10 ≐ **£233**
(to the nearest pound)
*(1 mark for 17 × £9.30,
1 mark for £158.10 + £75,
1 mark for the correct
answer)*

Test 5 — pages 10 and 11

1 10% of 300 = 30
80% of 300 = 30 × 8
= **240** *(1 mark)*

2 7 ÷ 3 = 2 remainder 1
= $2\frac{1}{3}$ *(1 mark)*

3 24 − 2 × 3 = 24 − 6
= **18** *(1 mark)*
You need to do the multiplication
before the subtraction.

4

$$
\begin{array}{r}
1\ 2\ 8 \\
\times\ \ \ 1\ 4 \\
\hline
5\,_1 1\,_3 2 \\
+\ 1\ 2\ 8\ 0 \\
\hline
1\ 7\ 9\ 2
\end{array}
$$

So £1.28 × 14 = **£17.92**.
*(1 mark for a correct method,
1 mark for the correct
answer)*

5

$$
\begin{array}{r}
{}^{9}\ {}^{9} \\
\cancel{5}\ \cancel{8}\ \cancel{0}\ \cancel{0}\ 0 \\
-3\ 6\ 7\ 5\ 0 \\
\hline
2\ 1\ 2\ 5\ 0
\end{array}
$$

Dividing by 5 will be easier
than dividing by 500:

$$
\begin{array}{r}
0\ 4\ 2\ 5\ 0 \\
5\overline{)\ 2\,^2 1\,^1 2\,^2 5\ 0}
\end{array}
$$

500 is 100 times greater than 5,
so now divide by 100:
4250 ÷ 100 = 42.5
So each smaller organisation
gets **£42.50**.
*(1 mark for a correct
subtraction, 1 mark for a
correct method of division,
1 mark for the correct
answer)*

Test 6 — pages 12 and 13

1 **0.3 0.28 0.24 0.219** *(1 mark)*

2 $\frac{6}{10}$ *(1 mark)*

3 15% of 4350 = $\frac{15}{100}$ × 4350
= **652.5** *(1 mark)*

4 3 hours at her desk
corresponds to 30 minutes
= 0.5 hours in meetings.
3 hours = 6 × 0.5 hours,
so multiply by 6 to convert
from meetings to desk.
3 hours 15 minutes = 3.25 hours
6 × 3.25 hours = **19.5 hours**
at her desk.
*(1 mark for correctly working
with the direct proportion,
1 mark for the correct
answer)*

5 Total rent = £30 247
4 + 3 = 7, so there are 7 parts.
1 part = £30 247 ÷ 7 = £4321
4 parts = £4321 × 4 = **£17 284**
*(1 mark for using 30 247,
1 mark for attempting to
use the ratio, 1 mark for the
correct answer)*

Test 7 — pages 14 and 15

1

$$
\begin{array}{r}
0\ 5\ 8 \\
6\overline{)\ 3\,^3 4\,^4 8}
\end{array}
$$

So 34.8 ÷ 6 = **5.8** *(1 mark)*

2 **31 200** *(1 mark)*

3 **7** *(1 mark)*

4 10% of 40 = 4 , 5% of 40 = 2
So 65% of 40
= 6 × 10% + 5%
= 6 × 4 + 2
= 24 + 2 = **26 buses**
*(1 mark for a correct method,
1 mark for the correct
answer)*

5 Centre A:
$\frac{1}{3}$ × 105 = 105 ÷ 3 = 35
Centre B:
$\frac{2}{5}$ × 105 = 105 ÷ 5 × 2 = 42
Difference:
42 − 35 = **7 animals**
*(1 mark for finding the
number of animals for
Centre A, 1 mark for finding
the number of animals for
Centre B, 1 mark for a correct
answer)*

Test 8 — pages 16 and 17

1 15^2 = 15 × 15 = **225** *(1 mark)*

2 $\frac{2}{7}$ × 6328 = **1808** *(1 mark)*

3 3 × 8 + 1 = 24 + 1 = 25,
so $3\frac{1}{8}$ = $\frac{25}{8}$. *(1 mark)*

4 8.5 miles ÷ 5 = 1.7
1.7 × 8 = 13.6 km, which is
greater than 12 km.
Yes — she is correct.
*(1 mark for one correct step
of the formula, 1 mark for
the correct conclusion with
working)*

5 a) 1 part = £1050
  There are 1 + 5 = 6 parts in
  total, so the total cost of the
  car is 6 × £1050 = **£6300**.
  *(1 mark for identifying
  £1050 is a sixth of the
  total, 1 mark for the
  correct answer)*

 b) $\frac{18}{25} = \frac{72}{100}$ = **72%** *(1 mark)*

Test 9 — pages 18 and 19

1 $\frac{1}{5}$ *(1 mark)*

  *Divide the top and the bottom of
  the fraction by 8.*

2 **0.178** *(1 mark)*

3   2 0
   4⟌8 0
  So 8 ÷ 0.4 = **20** *(1 mark)*

4 The two cities with the largest
  populations are Whereton and
  Whyton.

    1 3 3 9 2 0
   + 1 3 3 6 2 7
    ‾‾‾‾‾‾‾‾‾‾‾
    2 6 7 5 4 7
       1

  *(1 mark for using the
  populations of the correct two
  cities, 1 mark for the correct
  answer)*

5 a) 100 pies = 3000 g
   10 pies = 300 g
   70 pies = 7 × 300 = **2100 g**
   *(1 mark for working with
   the proportion, 1 mark for
   the correct answer)*

 b) 795.817 g ≈ 800 g
  800 g ÷ 4 = **200 g** *(1 mark)*

Test 10 — pages 20 and 21

1 **four hundred and five
  thousand, six hundred and
  twenty-eight** *(1 mark)*

2 Y = −1 and Z = 20
  20 − −1 = 20 + 1 = **21** *(1 mark)*

3 **40.7** *(1 mark)*

4 Increase = 5% of £23
     = $\frac{5}{100}$ × £23 = £1.15
  So the new price is
  £23 + £1.15 = **£24.15**.
  *(1 mark for a correct method,
  1 mark for the correct answer)*

5 40 × 600
   = 24 000 staplers per journey.
  180 000 ÷ 24 000 = 7.5, so
  8 journeys are needed.
  The truck needs to travel to
  the new building and back
  again, so each journey is
  12.5 × 2 = 25 miles.
  8 × 25 = **200 miles**
  *(1 mark for correctly using
  any two of 180 000, 40
  or 600, 1 mark for finding
  8 journeys, 1 mark for the
  correct answer)*
  *Alternatively, you could find the
  number of trolley loads needed
  (180 000 ÷ 600 = 300),
  then work out how many journeys
  would transport all of the trolley
  loads (300 ÷ 40 = 7.5).*

Section Two: Measures, Shape and Space

Test 1 — pages 22 and 23

1

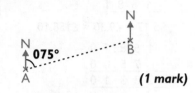

  (1 mark)

2 **rectangle** *(1 mark)*

3 £1389.60 − £677.80 − £379.99
   = **£331.81** *(1 mark)*

4 6.2 × 6.2 × 3.5 = **134.54 cm³**
  *(1 mark for the correct
  answer, 1 mark for the
  correct units)*

5 a) 5 cm = 1 km, so
   12 km = 12 × 5
      = **60 cm** *(1 mark)*

 b) Total time
   = $3\frac{1}{2}$ hours + 35 mins
       + 2 hours 50 minutes
   = $5\frac{1}{2}$ hours + 85 minutes
   = 6 hours 55 minutes
  08:30 + 6 hours 55 minutes
   = **15:25 or 3:25 pm**
  *(1 mark for the correct
  total time, 1 mark for the
  correct answer)*
  *Alternatively, you could have
  added on the time for each
  activity in turn to reach the
  same answer and still get the
  marks.*

Test 2 — pages 24 and 25

1 7530 ÷ 1000
   = **7.53 kg** *(1 mark)*

2 10% = £38 ÷ 10 = £3.80
  £38 − £3.80 = **£34.20** *(1 mark)*

3 09:10 → 13:10 = 4 hours
  13:10 → 13:42 = 32 minutes
  So Alun was on the train for
  4 hours, 32 minutes *(1 mark)*

4 E.g.

(1 mark for a quadrilateral with two pairs of parallel sides, 1 mark for a quadrilateral with no lines of symmetry)
The shape is a parallelogram.

5 Space within wooden border:
 height = 28 − 2 − 2 = 24 cm
 width = 24 − 2 − 2 = 20 cm

So the area within the border is 20 × 24 = 480 cm². Then the area not taken up by the photo is 480 − 150 = **330 cm²**.
(1 mark for using 24 and 20, 1 mark for finding the area within the wooden border, 1 mark for the correct answer)

Test 3 — pages 26 and 27

1 side elevation *(1 mark)*

2 reflex *(1 mark)*

3 13 × 13 × 13
 = **2197 cm³** *(1 mark)*

4 35% of £28.80 = £10.08
 Sale price = £28.80 − £10.08
 = £18.72
 No — she can't afford the plates.
 (1 mark for a correct method for finding the percentage, 1 mark for the correct conclusion with working)

5 Area of whole room
 = 10 × 7.2 = 72 m²
 Area of kitchen
 = 5 × 5 = 25 m²
 Area of carpet
 = 72 − 25 = 47 m²
 Number of ml = 22 × 47
 = **1034 ml**
 (1 mark for the correct method of finding the area of a rectangle, 1 mark for finding the area of the carpet, 1 mark for the correct answer)

Test 4 — pages 28 and 29

1 trapezium *(1 mark)*

2 360° *(1 mark)*

3 7 × 10 = **70 m** *(1 mark)*

4 7.8 m = 780 cm
 There are 780 ÷ 10 = 78 lots of 10 cm in 780 cm.
 £4.99 × 78 = **£389.22**
 (1 mark for converting the units correctly, 1 mark for the correct answer)

5 Split the plan into two rectangles:

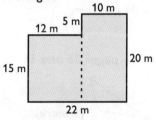

12 × 15 = 180 m²
10 × 20 = 200 m²
Area of floor = 180 + 200
 = **380 m²**
(1 mark for considering two rectangle areas, 1 mark for finding the area of one rectangle, 1 mark for the correct final answer)
You could have also split the shape horizontally and worked out 5 × 10 and 15 × 22.

Test 5 — pages 30 and 31

1 equilateral *(1 mark)*

2

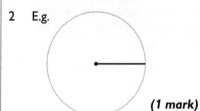

(1 mark)

3 5% of £1650 = $\frac{5}{100}$ × 1650
 = 82.5
 So she will owe **£82.50**. *(1 mark)*
 You will need to write £82.50 to get the marks, not just £82.5.

4 £19 − £6.24 = £12.76 profit on each box that he sells.
 5 boxes: 5 × 12.76 = 63.8
 So his profit will be **£63.80**.
 (1 mark for finding the profit for one box, 1 mark for the correct answer)

5 Volume of water butt
 = 2 m × 0.6 m × 0.5 m
 = 200 cm × 60 cm × 50 cm
 = 600 000 cm³
 600 000 ÷ 5000 = **120 times**
 (1 mark for converting the units, 1 mark for finding the volume of the water butt, 1 mark for the correct answer)

Test 6 — pages 32 and 33

1 cuboid *(1 mark)*

2 E.g.

(1 mark)

3   Cost of truffles
    = £1.82 × 6 = £10.92
    Balance left on card
    = £20 − £10.92
    = **£9.08 (1 mark)**

4   Perimeter of one card:
    = 11.2 + 8.7 + 11.2 + 8.7
    = 39.8 cm, so she needs
    39.8 cm of ribbon for one card.
    For 10 cards, she needs
    39.8 cm × 10 = **398 cm**.
    (1 mark for finding the ribbon needed for one card, 1 mark for the correct answer)

5   Boat:
     4 cm on the map,
     so 4 × 2.5 = 10 km.
    Train:
     1 + 3 + 2 + 1 = 7 cm on the map, so 7 × 2.5 = 17.5 km.
    So the journey by boat is
    17.5 − 10 = **7.5 km** longer.
    (1 mark for finding the journey by boat, 1 mark for finding the journey by train, 1 mark for the correct answer)
    You could also work out the difference in centimetres and convert this to kilometres at the end.

Test 7 — pages 34 and 35

1   **one (1 mark)**

2   $45\% = \frac{45}{100} \times £250 = £112.50$.
    So the balance will be
    £250 + £112.50 = **£362.50**
    (1 mark)

3   7 − 1.5 = **5.5 m (1 mark)**

4   E.g.

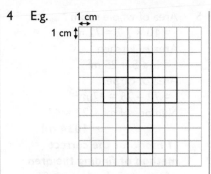

(3 marks for a correct net. Otherwise, 2 marks for six faces of the correct size which don't form a correct net, or five faces that create the net of an open cube, or the net of a cube which is the wrong size. 1 mark for at least one face drawn the correct size)

5   He needs 0.04 × 8 = 0.32 litres. The container has 0.3 litres left in it. **No** — he does not have enough detergent.
    (1 mark for using the correct container measurement, 1 mark for the correct conclusion with working)
    You could also divide 0.3 litres by 0.04 litres to see how many loads could be done with the amount left in the container.

Test 8 — pages 36 and 37

1

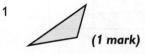

    (1 mark)

2   336 ÷ 12 ÷ 7 = **4 cm (1 mark)**

3   1920 ÷ 60
     = **32 hours (1 mark)**

4

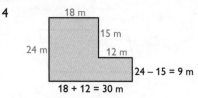

    Perimeter
    = 18 + 15 + 12 + 9 + 30 + 24
    = **108 m**
    (1 mark for finding the missing sides, 1 mark for the correct answer)

5   Phones R Gr8:
     20% of £238 = $\frac{20}{100} \times £238$
             = £47.60
    In total, Marvin would pay
    £238 + £47.60 = £285.60.
    I Heart Phones:
    £2.50 × 12 months = £30
    In total, Marvin would pay
    £263 + £30 = £293.
    No — I Heart Phone does not offer him the cheapest deal.
    (1 mark for finding the interest with Phones R Gr8, 1 mark for finding the charges with I Heart Phones, 1 mark for the correct conclusion with working)

Test 9 — 38 and 39

1   **135° (1 mark)**

2   **the diameter (1 mark)**

3   48 ÷ 4 = **12 cm (1 mark)**

4   Cost of shirt and headphones
     = £14.25 + £109 = £123.25
    Total amount spent
     = £435 − £239.25 = £195.75
    Cost of coffee machine
     = £195.75 − £123.25 = **£72.50**
    (1 mark for finding one of the amounts above, 1 mark for the correct answer)

5 7.30 pm – 30 minutes = 7.00 pm
7.00 pm – 45 minutes = 6.15 pm
So he needs to be in London
by 6.15 pm = 18:15.
The last train to arrive before
18:15 is the 1803, which leaves
York at **1600**.
*(1 mark for finding when
he should be in London by,
1 mark for choosing the train
that arrives at 1803, 1 mark
for the correct answer)*

Section Three: Handling Data

Test 1 — pages 40 and 41

1 **Unlikely** *(1 mark)*

2
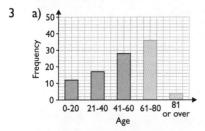

So the range is **77 974 tickets**.
*(1 mark for subtracting the
lowest value from the highest
value, 1 mark for the correct
answer)*

3 a)

*(2 marks for all three bars
correct, otherwise 1 mark
for two bars correct)*

b) 28 + 36 + 4
= **68 people** *(1 mark)*

4 5 + 9 = 14, so 14 of the prizes
are not notebooks.
So the probability of not
picking a notebook = $\frac{14}{20}$
*(1 mark for using 14, 1 mark
for the correct answer)*
Any equivalent fraction (e.g. $\frac{7}{10}$)
would also get the marks.

Test 2 — pages 42 and 43

1 477 – 301 = **176** *(1 mark)*

2 a)

	Tally	Frequency
Nails	卌 IIII	9
Screws	卌 I	6
Bolts	IIII	4
Total:		19

*(1 mark for a fully
correct table)*

b) There are 6 screws out of 19
pieces, so the probability
of picking out a screw = $\frac{6}{19}$.
(1 mark)

3 a) **25%** *(1 mark)*
The angle is 90°, which is a
quarter of the circle.

b) 360° ÷ 120 = **3°**
*(1 mark for dividing total
number of degrees by 120,
1 mark for the correct
answer)*

c) Monday: 39° ÷ 3 = 13 people
Tuesday: 72° ÷ 3 = 24 people
24 – 13 = **11**
*(1 mark for finding the
number of visitors on
Monday or Tuesday,
1 mark for the correct
answer)*

Test 3 — pages 44 and 45

1 **A card picked at random
from a set labelled 1 to 20 is
odd.** *(1 mark)*

2 E.g.

Number of boxes on pallet	Tally	Number of pallets
1-5		
6-10		
11-15		
16-20		
Total:		

*(2 marks for a fully correct
table with at least three
groups, otherwise 1 mark for
a grouped frequency table
with no more than one error)*
Examples of errors would include
not covering all the possible
numbers of boxes in your groups,
or having groups that overlap
(e.g. 1-5 and 5-10) You'd still
get the marks if you didn't draw a
totals row or tally column.

3 a) 228 ÷ 76 = 3

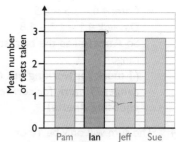

*(1 mark for finding the
mean, 1 mark for correct
bar as above)*

b) **No** — it's only true that
the average number of tests
taken is half, not that each of
Jeff's students has taken half
the number of tests as Sue's
students. *(1 mark)*

c) 396 ÷ 1.8 = **220 students**
*(1 mark for reading 1.8
from the bar chart, 1 mark
for the correct answer)*

Test 4 — pages 46 and 47

1 18 + 66 + 53 + 14 + 39 = 190
Mean = 190 ÷ 5 = **38** *(1 mark)*

2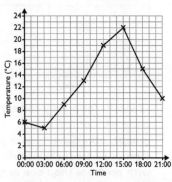

(1 mark for at least six points plotted correctly, 1 mark for joining all the plotted points with straight lines)

3 a) 3.4 – 2.5
= **0.9 points** *(1 mark)*

b) Anette: 4.0 – 2.0 = 2.0
Bernard: 3.7 – 3.1 = 0.6
Cate: 4.0 – 3.5 = 0.5
Daveed: 4.1 – 2.3 = 1.8
So **Anette** had the biggest range.
(1 mark for working out the ranges, 1 mark for correctly choosing Anette)

4 The sum will only be 10 or more if she rolls a 5 or 6 on the second dice. There are six possible outcomes of the second dice roll, so the probability is $\frac{2}{6}$.
(1 mark for finding the number of ways to get 10 or more, 1 mark for the correct answer)
You could simplify your answer to $\frac{1}{3}$.

Test 5 — pages 48 and 49

1 8 ÷ 20 = 0.4
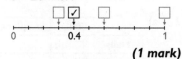
(1 mark)

2 Total number of cancellations
= 130 + 30 + 220 + 340 = 720
Trespass:
30 ÷ 720 × 360° = 15°
Signal failure:
220 ÷ 720 × 360° = 110°
Weather:
340 ÷ 720 × 360° = 170°

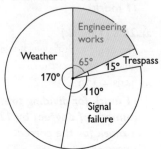

(3 marks for a fully correct, labelled pie chart, otherwise 2 marks for at least one angle worked out correctly and drawn, or 1 mark for starting to work with proportions)
Alternatively, you could use the angle for engineering works that you're given. 65° = 130 ÷ 2, so each angle is just half the corresponding number of cancellations.

3 a) **£2.50** *(1 mark)*

b) 2015: +£1.25
2017: –£0.50
2019: +£0.75
£1.25 – £0.50 + £0.75
= £1.50
Mean = £1.50 ÷ 3 = **£0.50**
(1 mark for adding £1.25, –£0.50 and £0.75, 1 mark for the correct answer)

c) **No** — the graph shows the differences in price between 2013 and 2019, not the actual prices. *(1 mark)*

Test 6 — pages 50 and 51

1 $\frac{1}{5}$ of 10 = 10 ÷ 5 = **2** *(1 mark)*

2 a) 22 – 4 = **18 hours** *(1 mark)*

b) **Spring** *(1 mark)*

3

Price	Frequency
£8.50 or less	2
£8.51–£9.50	9
£9.51–£10.50	6
£10.51 or over	4

(2 marks for all correct, otherwise 1 mark for at least two correct)

4 a) 8220 + 8457 + 7950 + 8065
= 32 692
Mean for Airline A
= 32 692 ÷ 4 = 8173
8148 is smaller than 8173, so **Airline B** had fewer minutes of delays on average.
(1 mark for attempting to find the mean for Airline A, 1 mark for the correct answer with explanation)

b) 7879 + 362 = **8241** *(1 mark)*

Section Four:
Mixed Practice

Test 1 — pages 52 and 53

1   1000 + 100 450 ÷ 1000
    = 1000 + 100.45
    = **1100.45 (1 mark)**
    You have to do the division
    before the addition.

2   Ada: 75% = 0.75, Ben: $\frac{4}{5}$ = 0.8
    In order from highest to lowest:
    Ben, Ada, Cal (1 mark)

3   330 ÷ 3 = 110
    110 × 2 = **220 ml (1 mark)**

4   Biggest = 82.95, smallest = 35.88

$$\begin{array}{r} {}^{7}\cancel{8}\,{}^{12}\cancel{2}.\,{}^{8}\cancel{9}\,{}^{15}\cancel{5} \\ -\ 3\ 5\ .\ 8\ 8 \\ \hline 4\ 7\ .\ 0\ 7 \end{array}$$

   **(1 mark for 82.95 – 35.88,
   1 mark for the correct
   answer)**

5   Width + length
    = 16.8 ÷ 2 = 8.4 m
    So 3 + 4 = 7 parts = 8.4 m
    and 1 part = 8.4 ÷ 7 = 1.2 m.
    So the width of the rectangle is
    4 parts = 1.2 × 4 = **4.8 m.**
    **(1 mark for using 8.4 m,
    1 mark for working correctly
    with the ratio, 1 mark for the
    correct answer)**
    Alternatively, you could use the
    ratio with the whole perimeter and
    find that the width (both top and
    bottom) contributes 9.6 m of the
    perimeter. You then need to divide
    by 2 at the end.

Test 2 — pages 54 and 55

1   **4 (1 mark)**

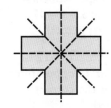

2   5% = $\frac{5}{100}$ × £206 250
    = £10 312.50
    Price = £206 250 – £10 312.50
    = £195 937.50
    **One hundred and ninety-
    five thousand, nine hundred
    and thirty-seven pounds
    and fifty pence (1 mark)**

3   2 × 4 + 3 = 11, so $\frac{11}{4}$ **(1 mark)**

4   The pet needs an injection
    at the start of the first year,
    then at the start of the fourth,
    seventh and tenth. So 4 are
    needed in total.
    Total cost
    = £63.98 + £32.60 × 4
    = **£194.38**
    **(1 mark for £32.60 × 4,
    1 mark for the correct
    answer)**

5   Summer mean
    = (26 + 24 + 16) ÷ 3 = 22
    Autumn mean
    = (14 + 14 + 11) ÷ 3 = 13

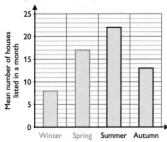

   **(3 marks for a fully correct
   bar chart, otherwise 2 marks
   for one bar drawn at the
   correct height, or 1 mark for
   evidence of a correct method
   to find a mean)**

Test 3 — pages 56 and 57

1   Angle of shaded region = 120°
    Angle of full circle = 360°
    So $\frac{120}{360}$ **(1 mark)**

2   **Net C (1 mark)**

3   Area = height × width
    = 3.94 × 10.26
    ≈ 4 × 10
    = **40 cm² (1 mark)**

4   Red roses are 2 parts out of
    5 parts, so the probability
    is $\frac{2}{5}$ = 0.4.

   (1 mark)

5   a)

Weight	Tally	Frequency
500-524	II	2
525-549	ЖИ II	7
550-574	ЖИ III	8
574-600	III	3
	Total	20

   **(1 mark for each correct
   row)**

   b) Heaviest bag = 590 g
   10% = 590 ÷ 10 = 59 g
   590 + 59 = **649 g**
   **(1 mark for finding that
   10% = 59 g, 1 mark for
   the correct answer)**

Test 4 — pages 58 and 59

1   1000 – –995 = 1000 + 995
    = **1995 (1 mark)**

2   3 × 2 = 6 m²
    6 × 4 = 24 m²

    6 + 24 = **30 m² (1 mark)**

3 1 cm represents 100 cm = 1 m,
so 12 cm represents **12 m**.
(1 mark)

4 $15\% = \frac{15}{100} \times £1299 = £194.85$
Total owed = £1299 + £194.85
= **£1493.85**
*(1 mark for 15% = £194.85,
1 mark for the correct
answer)*

5 Angle for 1 response
= 360° ÷ 120° = 3°
Angle for Agree = 3° × 25 = 75°
Angle for Disagree
= 3° × 65 = 195°

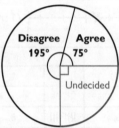

*(1 mark for working with
proportions of 360°,
1 mark for at least one angle
correctly calculated,
1 mark for a fully correct,
labelled pie chart)*
You could work out the third angle
by subtracting the two angles you
know from 360°.

Test 5 — pages 60 and 61

1 $\frac{3}{10}$ and $\frac{2}{8}$ *(1 mark)*

2 1.5 × 2.3 × 2.6
= **8.97 cm³** *(1 mark)*

3 $30\% = \frac{30}{100} \times 14$
= **4.2 minutes** *(1 mark)*

4 9 am → 1 pm is 4 hours.
1 hour = 3 × 20 minutes, so
there are 3 appointments
per hour and 3 × 4 = 12
appointments in total.
So the probability is $\frac{1}{12}$.
*(1 mark for 12 appointments,
1 mark for the correct
probability)*

5 Width of lit area
= 100 ÷ (8.96 + 64)
= 1.3706.... m
= **1.37 m** (to 2 d.p.)
*(1 mark for putting the
numbers into the formula
correctly, 1 mark for
calculating 1.3706..., 1 mark
for the correct rounded
answer)*